1. Water Is Important 5

2. The Water Cycle Can Affect Sources 8

3. What Causes a Water Shortage? 11

4. Water Is a Basic Right 14

5. People Travel a Long Way for Clean Water 17

6. Unsafe Water Increases Health and Safety Problems 21

7. Demand for Water Is Growing 24

8. Pollution Makes Water Unsafe 27

9. Cleaning Water Is Hard and Costly 30

10. Moving Water Is Also Costly 34

11. Climate Change Affects Water Access 37

12. Countries Must Work Together to Create Solutions 40

How Can You Help? 44

Glossary 46

For More Information 47

Index 48

Dams and reservoirs help to provide a stable water supply.

TODAY'S HEADLINES

12 THINGS TO KNOW ABOUT WATER ACCESS

KRISTIN EBERTH

BLACK RABBIT BOOKS

Table of Contents

Water Is *Important*

1

Water is important for all living things. It is the most important liquid in the world. Our bodies are made up mostly of water. People, plants, and animals all need water to survive. Some organisms are made up of 90 percent water.

Our planet is mostly covered in water. It may seem like water is plentiful. But most of this is saltwater. It is too salty to drink. Only 3 percent of the Earth's water is fresh water. This is the water we can safely drink and use.

People get their water from many different sources. In some parts of the world, people get their water from rain. They use tarps and rooftops to collect it. Then they store it in large containers. Rivers, lakes, and streams are common sources too. **Reservoirs** are large lakes created by dams. They store large amounts of water for later use. Towns and cities are usually near a source of water. This water must be treated or cleaned before it can be safely used. Groundwater is another source. Some people dig wells to get this water. These are deep holes in the ground. People pump water up from **aquifers**.

A diagram shows how people get water from an aquifer.

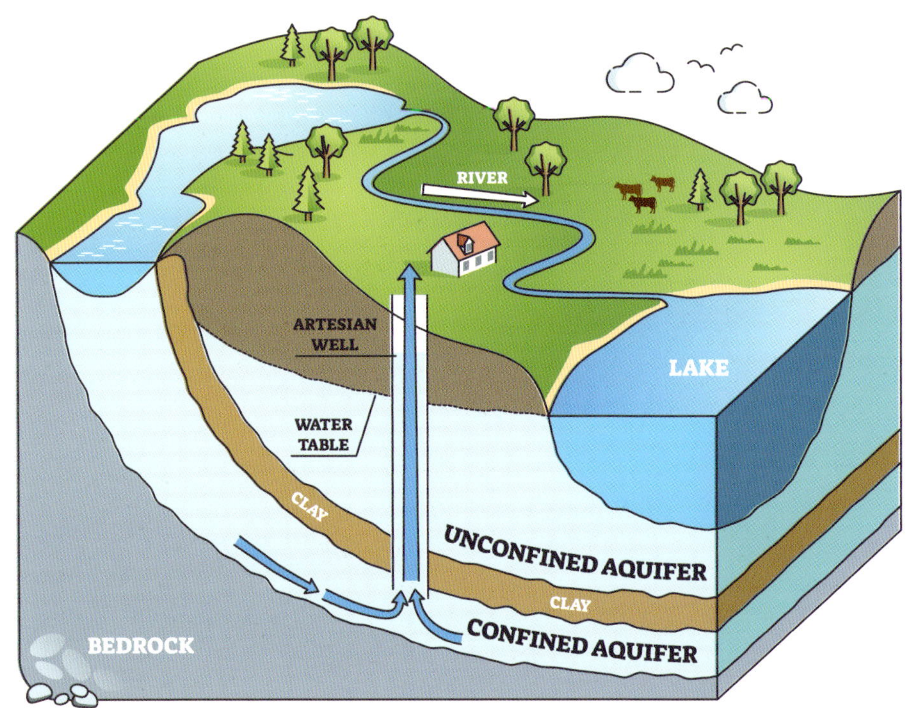

0.5 Percentage of the Earth's fresh water that is available.

Most fresh water is frozen in glaciers and polar ice caps. • Some fresh water lies too far below the surface to be dug out. • Some sources are too polluted to use.

Think About It

Where does your water come from?

People collect rain in barrels to use in gardens and landscaping.

The Water Cycle *Can Affect* Sources

2

Have you ever wondered where the water comes from? It is all part of the water cycle.

The water cycle is nature's way of recycling water. It starts at a water source. The sun heats up the water. This causes it to turn into water vapor. This process is called **evaporation**. Once the vapor is in the air, it cools down. It turns into tiny droplets of water. These droplets form clouds in a process called **condensation**. When the drops get too heavy, they fall. They go back down to the ground as rain or snow. This is called **precipitation**.

Reduce, reuse, recycle can apply to waste and water.

When it rains or snow melts, the water soaks into the land. This is groundwater. It helps plants grow. The water moves through plants and evaporates. Some water flows back into lakes and rivers. Then the cycle starts again.

96.5 Percentage of the Earth's water that is in oceans.

1.7 percent is in lakes, streams, rivers, and soil. • 0.001 percent is in the Earth's atmosphere. • 1.7 percent is in polar ice caps, glaciers, and permanent snow.

The water cycle is constantly repeating.

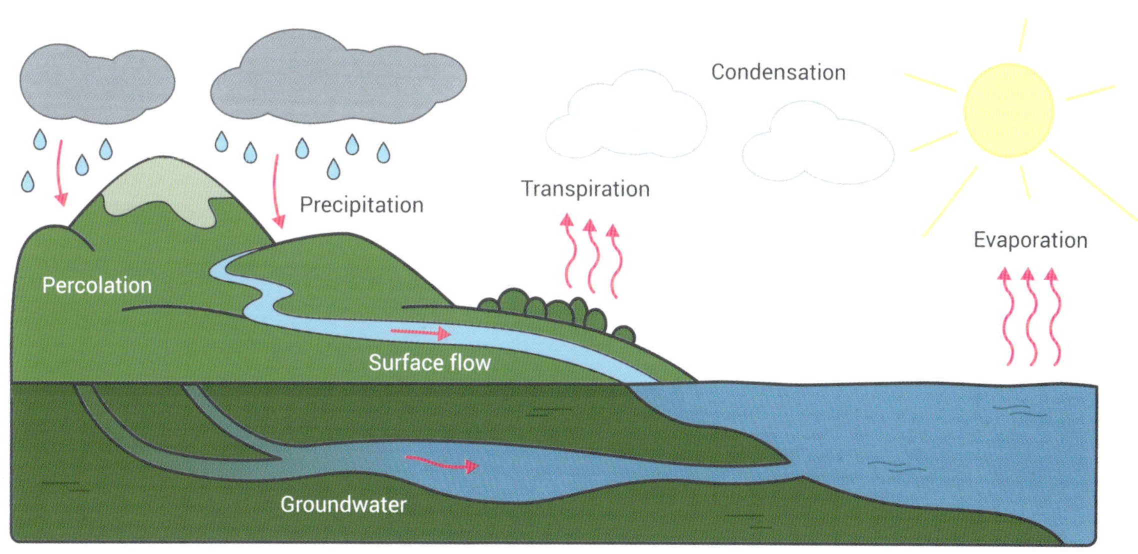

Rain doesn't always fall in the area where it evaporated. Clouds transport the droplets until they get too heavy. The process can take days. The clouds will have moved over some other place by then. The Earth never loses any water. It just changes form and location. If a place does not get enough rain, it can create a water shortage.

BIGGEST AND OLDEST LAKES

Lake Superior is in the northern United States. It is the world's largest freshwater lake by area. It is 31,700 square miles (82,103 square kilometers). It is also the deepest and coldest of the Great Lakes. Lake Baikal is located in Russia. It is the oldest freshwater lake. It has been dated back to 25 million years ago.

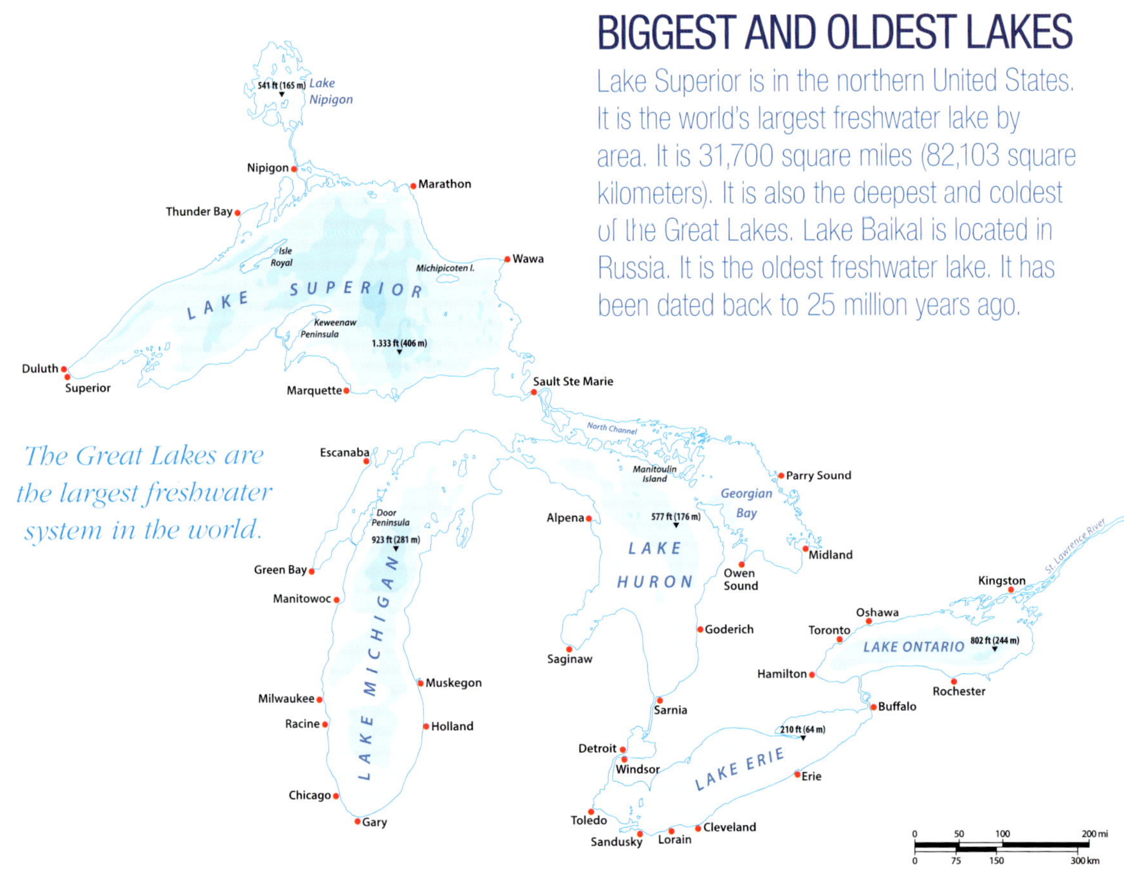

The Great Lakes are the largest freshwater system in the world.

What Causes a Water *Shortage?*

3

Imagine you wake up thirsty. You turn on the faucet, but nothing comes out. Not a drop of water. That is what happens during a water shortage. There isn't enough water for everyone to use.

Shortages happen all over the world for many different reasons. Pollution makes water unusable. Weather can damage sources of water. Groundwater can dry up when people use too much. If a population increases, more people need water. A water source may not be able to provide for more people. The water runs out faster.

When water is poorly managed, it can be wasted. City pipes can leak water, wasting it. If a city does not have a reliable way of getting water, people will have a shortage. Sometimes a shortage happens in a small area. Sometimes, a whole city or county is affected.

Droughts cause shortages too. A drought is when very little rain falls for a long time. The ground gets dry. Plants have a hard time growing. If the drought lasts a while, water sources run low. Creeks, lakes, and rivers can dry up. It's important to **conserve** water. This helps prevent shortages.

Water shortages are common in drier places, like deserts.

1 in 3 Number of schools that lack access to water globally.

Water is used for **sanitation**. Without it, viruses and disease can spread. • Without toilets, students must leave school to go to the bathroom. • Drinking water improves memory and focus.

Think About It

What would you do if water was in short supply?

Oil spills pollute the water and make it unsafe to use.

Water Is a *Basic Right*

4

Access to safe water is a basic human right. Access to water means people can get it when they want to. Billions of people around the world do not have this access. Safe water is clean water. It is free of any germs that could make you sick.

Water is needed for drinking and cooking. It is also needed for sanitation and **hygiene**. Poverty makes getting water harder. It is costly to build the **infrastructure** needed to get water. Many people in poverty do not have access. In the United States, over 2.2 million people live in homes without basic plumbing. Native Americans are 19 times more likely to lack indoor plumbing. Blacks and Latinos are twice as likely.

The United Nations (UN) created a goal in 2015. It hopes to provide water access for everyone by 2030.

Not all homes have safe drinking they can get from a faucet.

This will be clean drinking water. It will be affordable too. The UN also hopes to give access for toilets and washing. This is important for women and girls. Often, it is the women or children who collect water for homes. This can impact work and schooling. They are also at more risk for injury and danger.

A woman in India collects water from a community well.

People Travel a Long Way *for Clean Water*

5

Around the world, millions of people travel to get water every day. This can be hard and take a lot of time. They might have to walk for hours to reach a river or a well. The path they take may be dangerous. There could be bad weather, floods, or fighting nearby. When they get to the water, sometimes the well is locked up. Sometimes the water is polluted. If the water is safe, they must carry it all the way back home.

Water must be collected every day. Even if someone is sick, they must make the journey. Otherwise, there is no water to drink. It can be hard to carry enough water for everyone in the family. Farmers need a lot of water for their crops. Many people make multiple trips in a day. The trip is tiring. It is hard on a child's body. It also makes it hard to go to school.

Traveling long distances for water makes it hard to survive. Many organizations are helping. They dig new wells and build infrastructure. This brings water to areas that do not have access. It helps communities grow and thrive.

Crops need a lot of water to grow.

Think About It

How much water does your family use in a day? Could you carry that amount of water?

The bathroom is where the most water is used in an average household.

Unsafe Water Increases Health and *Safety Problems*

6

Not having clean water makes life much harder. It makes it hard to stay healthy. Polluted water can cause eye problems, skin infections, and diseases. Without water, it is hard to wash your hands. Germs spread and make people sick. Cholera, dysentery, and typhoid fever can spread through dirty water. They cause diarrhea, high fevers, and pain. People can die if they are not treated.

Your body cannot function right without water. Adults and children can die from lack of water and food. People use water to grow food. If there isn't enough, farmers can't water crops or raise animals. This can lead to a food shortage.

Water shortages create water **refugees**. These people leave their homes because there is no water. When faced with hunger and thirst, people may

Vaccines can help prevent some diseases.

FIGHTING TO SURVIVE People fight wars over water. In a drought, countries may not agree on how to share or use the little water there is. In some African countries, farmers and herders fight over water. People are desperate to save their families. In countries like Yemen, India, and Kenya, the fights have turned deadly.

People without access to clean water rely on water from a lake or river.

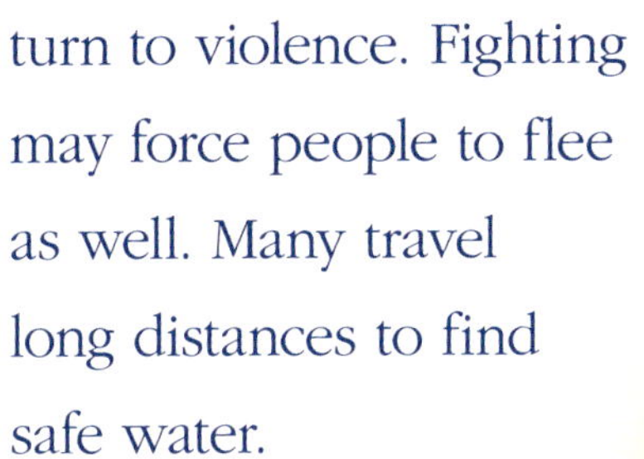

turn to violence. Fighting may force people to flee as well. Many travel long distances to find safe water.

40 million Number of people impacted by the drought across East Africa from 2020 to 2024.

This is the worst drought in 40 years. • There are 1.1 million refugees from this drought. • A 10 percent rise in global migration is linked to lack of water.

A girl in Africa transports jugs of water with a wheelbarrow.

Demand for Water *Is Growing*

7

The world's population is growing. More people on Earth means more water is needed. This makes our water sources even more important.

Water use grows at almost twice the rate of the population. In the United States, there are 204 freshwater basins. Nearly half of these may not meet monthly demands by 2071. By 2050, there will be about 9 million people on Earth. Agriculture will need to increase by 50 percent to feed everyone.

Agriculture uses the most water. It uses about 70 percent of the fresh water available. Livestock need water to live. Crops do too. A pound of wheat takes 130 gallons (492 liters) to grow. Coffee needs 2,500 gallons (9,464 L) per pound! Fields are usually **irrigated**. Water is pumped through pipes and sprayed

30 Number of gallons (114 L) of water an average family uses every week.

About 70 percent of this is used indoors. • Toilets use 24 percent of indoor water. • Leaks can waste about 13 percent of water.

A cow drinks about 12 gallons (45 L) of water every day.

over the fields. Some of this water runs off or evaporates. It is important farmers use water wisely.

Our choices affect water usage too. We can choose more eco-friendly options. A pound (0.45 kg) of beef takes 1,800 gallons (6,814 L) of water to produce. That's enough for 39 bathtubs. Eating vegetarian options can help save water.

Think About It

Research areas that have water shortages. What do you notice that these areas have in common?

Beef, coffee, and crops all take a lot of water to produce.

Pollution Makes *Water Unsafe*

8

Polluted water is unsafe to drink or swim in. Pollution can come from chemicals or waste. Sometimes harmful substances spill into a lake or river. Once the water is polluted, it can enter other waterways. It can seep into the groundwater.

Pollution isn't always from chemicals. Factories use water to cool their equipment. The water can become very hot. If it is released into a lake, it can raise the temperature of the lake water. This is called **thermal** pollution. This can damage ecosystems.

A sign warns people not dump waste into a water source to protect the ocean.

Pollution can come from rain. In cities, rain runs along the pavement. This is called **runoff**. It may pick up chemicals, waste, and heavy metals. Near farms, runoff picks up pesticides and fertilizer. This pollutes local lakes and

Water pollution also hurts animals and plants near the water.

rivers. This makes the water unsafe for drinking or growing food. It can also hurt the wildlife living near the water source.

A specific kind of runoff is **nutrient** pollution. This is when phosphorus or nitrogen get added to water. These help algae grow like crazy in ponds and lakes. Algae consumes oxygen. It blocks sunlight. Underwater plants can't grow. Fish who eat the plants may die out.

SMALL THINGS ADD UP

Oil spills are bad for water. About 1 million tons (907,187 metric tons) of oil pollute waters each year. Almost half of that comes from factories, farms, and cities. Oil and gas drips from millions of vehicles each day. Big spills can be cleaned up—mostly. The smaller spills can be more harmful. They are harder to stop.

Cleaning Water Is Hard *and Costly*

9

Cleaning polluted water takes a lot of technology and effort. The hardest part is figuring out what is making it dirty. Water pollution is caused by many things. It could be road salt, fertilizer, or animal waste.

Home filters remove the big particles, but they do not clean out bacteria.

One cleaning method is a **filter**. It catches bigger particles like pebbles, dirt, and waste. They can be removed easily. A fine mesh can catch even very small particles like sand.

Water can also be cleaned with chemicals. These remove bacteria and viruses from water. Some chemicals make the waste stick together. The larger clumps are easier to filter out of the water. Some chemicals, like chlorine, disinfect the water. These chemicals kill any viruses or bacteria in the water.

It is hard when water has more than one thing making it dirty. It requires many cleaning processes. This is costly. Each process uses a lot of energy and materials. In addition, water sources usually connect to each other. A river may run into other lakes or streams. If water is polluted upstream, the whole source may become unsafe. Everything the dirty water touches will need cleaning. Wildlife can be affected. It takes many miles (kilometers) of clean up.

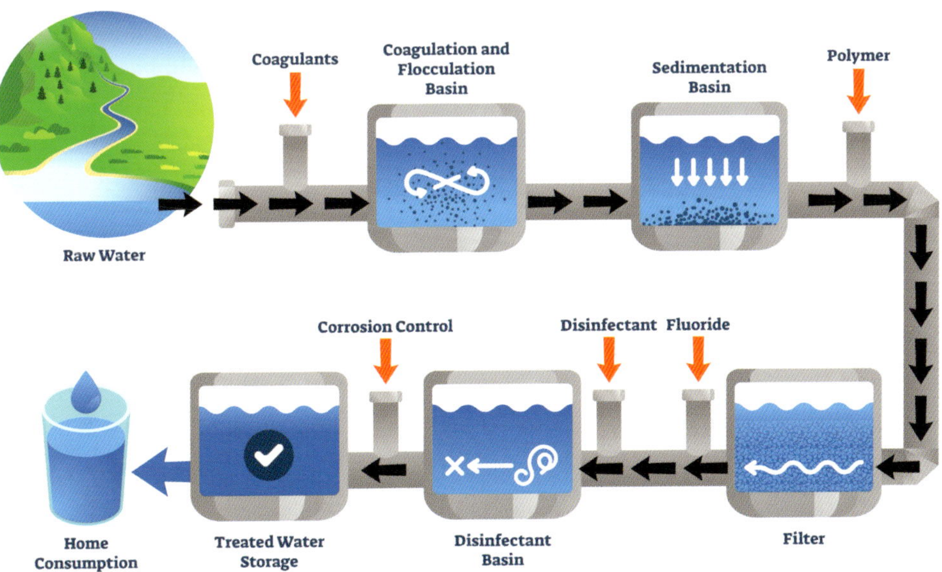

1.7 billion Number of people worldwide drinking water polluted with waste.

Unsafe water kills about 1,000 children under five each day. • About 818 million children don't have water or soap to wash their hands.

Wastewater treatment facilities work to clean water.

Think About It

Should water be free to everyone? Or should we pay more for water, so we don't waste it?

Moving Water Is Also *Costly*

10

If water isn't available in an area, it may be pumped in. This can help bring water to those in need. But it is a tricky and expensive solution. It takes a lot of energy, materials, and time.

The easiest way to move water is with big trucks. They carry tanks filled with water. The trucks drive to areas that need water and unload it. This doesn't work if a whole city is in need though.

Another way is to build new infrastructure. Pipelines pump water over long distances. They are usually underground. They reach places that are far away from a water source. Pipes are a convenient way to get water into homes. But they are difficult to run through mountains or over rivers. The construction is costly. It takes a lot of planning, time, and resources.

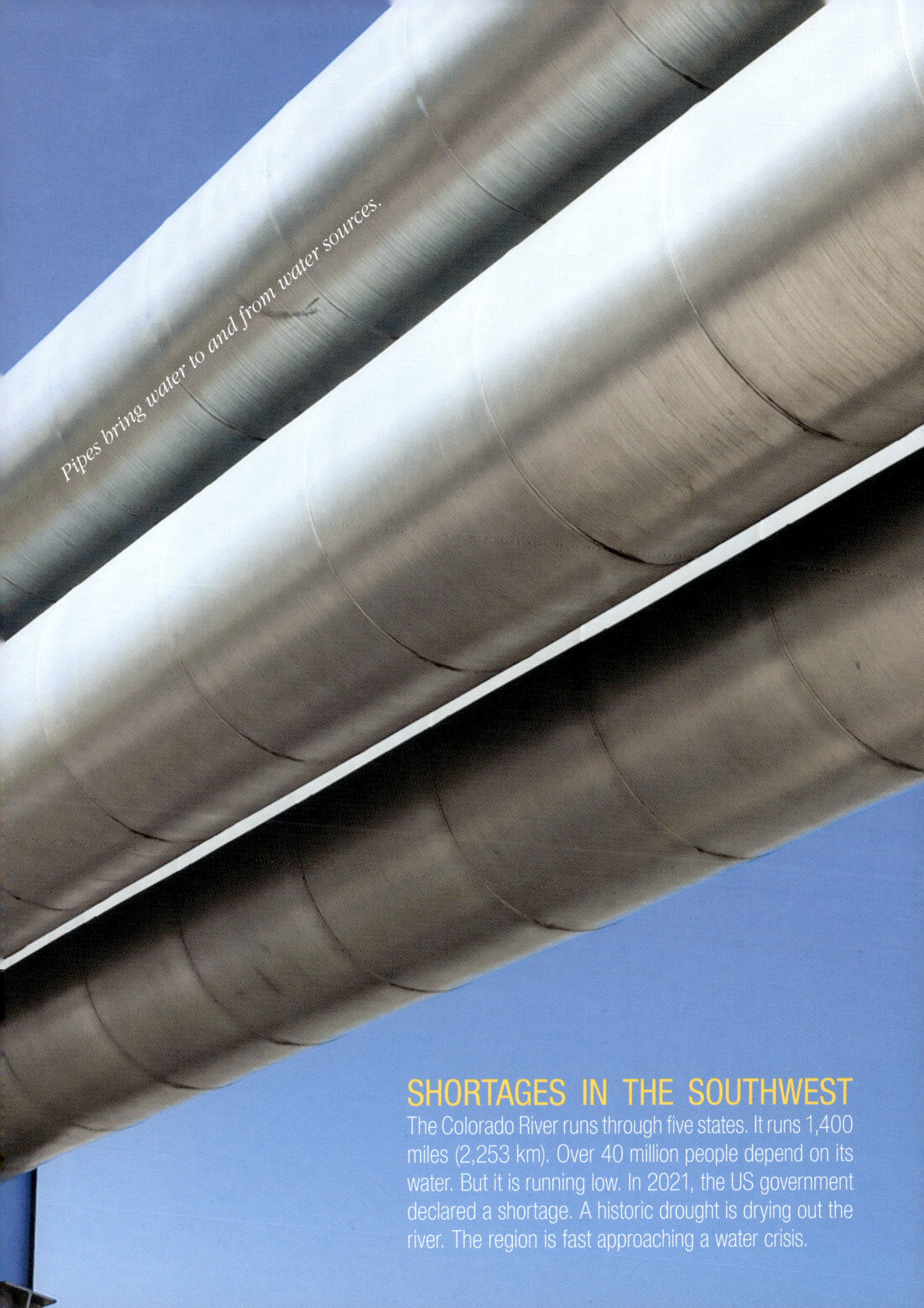

Pipes bring water to and from water sources.

SHORTAGES IN THE SOUTHWEST

The Colorado River runs through five states. It runs 1,400 miles (2,253 km). Over 40 million people depend on its water. But it is running low. In 2021, the US government declared a shortage. A historic drought is drying out the river. The region is fast approaching a water crisis.

An **aqueduct** is another human-made solution. This system uses ditches, canals, and tunnels to move water. Dams are used to store water. They create a reservoir that can be used for irrigation. They also can create electric power and control flooding.

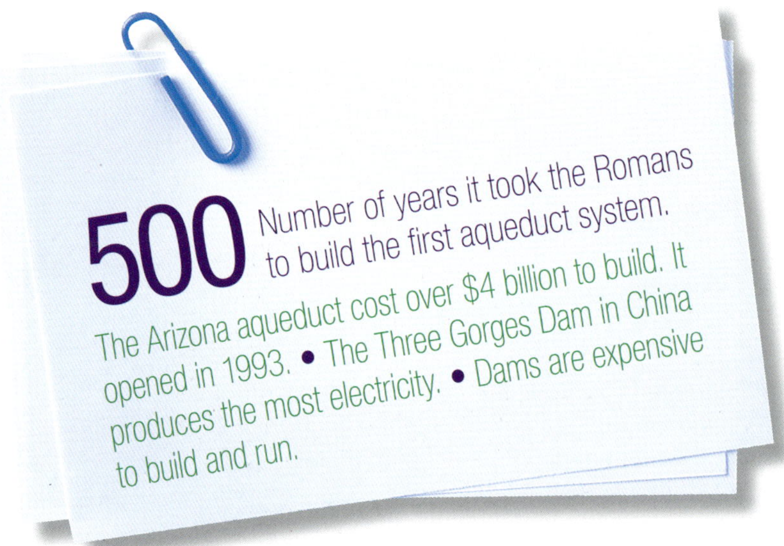

500 Number of years it took the Romans to build the first aqueduct system.

The Arizona aqueduct cost over $4 billion to build. It opened in 1993. • The Three Gorges Dam in China produces the most electricity. • Dams are expensive to build and run.

Climate Change Affects *Water Access*

11

Climate change is making the Earth's temperature rise. This change in temperature is causing our weather patterns to shift. Storms are bigger and stronger. Some areas are getting too much rain. Other areas are not getting any rain at all. This can lead to some big problems, including access to fresh water.

Climate change is expected to make water access worse. People in dry areas will have less fresh water to use. In other areas, too much rain can cause flooding. Floods create dirty runoff and pollute water sources. During hurricanes, the most dangerous thing are storm surges. In a storm surge, salty ocean water floods the streets and pollutes freshwater sources. It can take weeks to make the water safe to drink again. Global warming also makes ice melt. Glaciers near the poles melt into the ocean. This extra water will make sea levels rise all over the world. Salt water will mix with fresh water. This would pollute the freshwater sources permanently.

To help, we can fight for laws that reduce pollution. We can heat and cool our homes with solar or wind energy instead of fossil fuels. This can help slow down climate change and protect our fresh water.

Climate change is causing some places on Earth to get hotter and drier.

90 Percentage of scientists who believe climate change is caused by humans.

Over 37.4 billion tons of carbon dioxide were released into the atmosphere in 2023. • That is a 1.1 percent gain over 2022. • The Earth is now about 1.98 °F (1.1 °C) warmer than it was in the 1800s.

Rainfall from a hurricane causes flooding in Florida.

Countries Must Work Together to *Create Solutions*

12

Water access is a global problem. Many water sources cross national borders. Water can create peace or spark conflict, according to the UN. To avoid clashes, countries must work together to solve water problems.

About 60 percent of the world's fresh water flows between countries. This affects more than 3 billion people. Many countries have water treaties. These help secure water access even during war. The treaties also help prevent, control, and reduce water pollution. India and Pakistan share access to the Indus River system. The Nile River is the longest river in the world. It is shared by 10 countries, including Egypt, Sudan, and Uganda. The United States has agreements with Canada for the Great Lakes and Mexico for the Colorado River.

The Nile is located in the northeast part of Africa.

World Water Day is on March 22. Each year, the UN raises awareness of water shortages and inequality. Clean water and sanitation for all is one of 17 goals the UN has created for 2030. World leaders support this goal. In 2023, the United States pledged $49 billion to help. By cooperating with each other, countries can balance everyone's water needs for a better future.

The Great Lakes are shared by the United States and Canada.

It is important
to work together
to save the
world's water.
153
Number of countries that share
freshwater sources.
Only 24 countries have agreements for all the water
they share. • A source that crosses or shares a
country's border is called transboundary. • Actions
one country or state takes can affect water access
downstream. This is why agreements are important.

How Can

Conserve Water

Make sure you use water wisely. Fix leaks and take short showers. Turn off the faucet when it is not in use.

Reuse Water

Did you know you can recycle water? Place a bucket in your shower while it is warming up. Use that water for cleaning. Catch rain from your roof in a rain barrel. On dry days, you can use this to water your flowers.

Protect Water Sources

We need to keep rivers, lakes, and oceans clean. Don't pour toxic or hazardous chemicals down the drain. Volunteer to help clean up litter.

You Help?

Support Water Projects
Some organizations give clean water to people in need. Some are digging wells and creating new infrastructure. You can support them. Donate your time or money.

Use Efficient Appliances
Try to run the dishwasher and washing machines less. Only run them when they are full. You can also get faucets and toilets that use less water.

Raise Awareness
Teach others about the importance of clean water. Teach your friends how they can help conserve water.

Glossary

aqueduct
A structure that looks like a bridge that is used to carry water.

aquifer
A layer of rock or sand that can absorb and hold water.

condensation
The process of turning from gas to liquid.

conserve
To use carefully in order to prevent loss or waste.

drought
A long period of time during which there is very little or no rain.

evaporation
The process of turning from liquid to gas.

filter
A substance with pores through which a gas or liquid is passed in order to separate out floating matter.

hygiene
The things you do to keep yourself clean.

infrastructure
The basic equipment and structures that are needed for a country or region to function properly.

irrigate
To supply crops with water using human-made means.

nutrient
A substance or ingredient a person or animal needs to be healthy.

precipitation
Water that falls to the earth, such as rain or snow.

refugee
A person who flees their country for safety, usually because of war or for religious reasons.

reservoir
A human-made lake that is used to store a large supply of water.

runoff
Water from rain or snow that flows over the ground into streams.

sanitation
The process of keeping clean relating to public health.

thermal
Related to heat.

For More Information

Books

Barr, Catherine. *Water: How We Can Protect Our Freshwater.* Somerville, MA.: Candlewick Press, 2023.

Furgang, Kathy. *Clean Water and Our Future.* New York: PowerKids Press, 2022.

Knutson, Julie. *Do the Work! Clean Water and Sanitation.* Ann Arbor, MI: Cherry Lake Publishing, 2022.

Websites

Aquation: The Freshwater Access Game
ssec.si.edu/sites/default/files/games/Aquation/index.html

EPA: WaterSense for Kids
www.epa.gov/watersense/watersense-kids

Water and My World
kids.niehs.nih.gov/topics/pollution/water

About the Author

Kristin Eberth lives in northern Minnesota with her husband and two daughters. When she's away from her writing desk, Kristin can be found out in nature. She's always up for an adventure.

Index

agriculture, 24
aqueducts, 36

cleaning water, 30–33
climate change, 37–39
collecting water, 7, 15, 16, 17, 18

dams, 4, 6, 36
droughts, 12, 22, 23, 35

fresh water, 5, 7, 10, 24, 37, 38, 40

glaciers, 7, 9, 37
Great Lakes, 10, 40, 42
groundwater, 6, 8, 11, 27

health problems, 21

infrastructure, 14, 18, 34

pollution, 11, 13, 27, 28, 29, 30, 38, 40
poverty, 14

runoff, 27, 29, 37

salt water, 5, 37, 38
sanitation, 13, 14, 42
shortages, 10, 11, 12, 21, 35
solutions, 12, 40, 42, 44–45

treaties, 40, 43

United Nations (UN), 14, 16, 40, 42

violence, 22, 23

water cycle, 8, 9
world population, 11, 24
World Water Day, 42

 • Top Rank is an imprint of Black Rabbit Books. • Edited by Alissa Thielges | Designed by Danny Nanos • Photographs © Dreamstime/Andrew7726, 5, 43, Davidmartyn, 4, Htfimages, 44, Ivan Zelenin, 20, Multiart61, 45, Pakhnyushchyy, 8, Samrat35, 16, Shuen Ho Wang, 12, Vgraphic Farao, 7, VectorMine, 31, Weeranuch Phiokham, 17; Getty Images/Artur Debat, 38; Pexels/Gül Isik, 2-3; Shutterstock/ADragan, 29, Ajdin Kamber, 22, Andreas von Mallinckrodt, 44, Andrey_Popov, 45, Andromeda stock, 32–33, Angel DiBilio, 36, Anton Starikov, 18, Belinda Pretorius, 48, BEST-BACKGROUNDS, 41, 42, Bilanol, 39, chanasorn jele, cover, Christian Vinces, 13, Dennis MacDonald, 24, dios libre, 9, DOERS, 2, Francesca_Bluth, 34, fotokaleinar, 35, GSPstock, 30, innakreativ, 38, Jolygon, 46–47, Kitsawet Saethao, 21, Lukas Gojda, 8, 9, Lyudmila Lucienne, 28, MaraZe, 26, Muhammad Farhat Zahoor, 11, New Africa, 26, Nopparat Promtha, 26, O partime photo, 27, Pixel-Shot, 15, Rainer Lesniewski, 10, Riccardo Mayer, 23, Roberto Vivancos, 27, 44, SewCreamStudio, 43, Tan_supa, 37, VectorMine, 6, Vitaliy Andreev, 18–19, Vova Shevchuk, 14, William Edge, 25, Wor Jun, 45 • Printed in China

Library of Congress Cataloging-in-Publication Data Names: Eberth, Kristin, author. | Title: 12 things to know about water access / by Krissy Eberth. | Description: Mankato, MN: Top Rank, an imprint of Black Rabbit Books, [2025] | Series: Today's headlines | Includes bibliographical references and index. | Audience: Ages 9–13 | Audience: Grades 4–6 | Identifiers: LCCN 2024020335 | ISBN 9781645823902 (library binding) | ISBN 9781645824121 (paperback) | ISBN 9781645824343 (ebook) | Subjects: LCSH: Water-supply—Juvenile literature. | Classification: LCC TD348 .E24 2025 (print) | LCC TD348 (ebook) | DDC 363.6/1--dc23/eng/20240521 | LC record available at https://lccn.loc.gov/2024020335